5 31

ENERGY

written by
Julian Marshall
and
Steve Pollock

Illustrated by
Nick Shewring, Jon Sayer, Ross Watton and Mike Atkinson

Designed by
Peter Shirtcliffe

Edited by
Sarah Dann

Picture Research by
Helen Taylor

CONTENTS

Energy in action

If something happens, it happens because of energy. Everything on this page needed energy to happen.

Very big happenings

This space shuttle needs energy to get it up away from the earth. The energy is released by burning the fuel in the rocket launcher.

The energy for volcanic eruptions comes from the heat deep down in the earth.

Thunder and lightning release electrical energy in the sky which we can see and hear as light and sound.

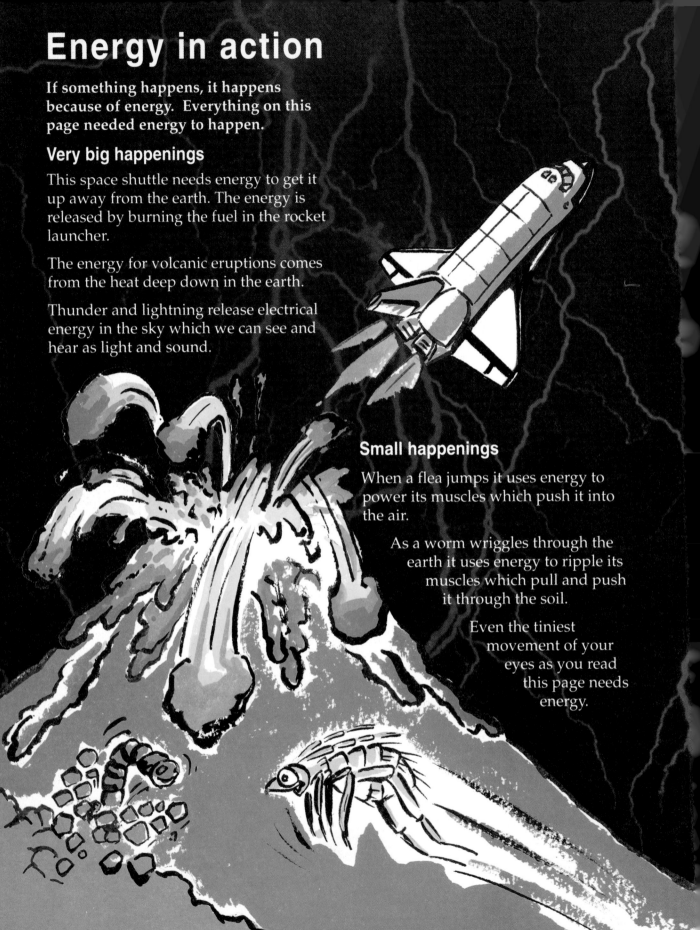

Small happenings

When a flea jumps it uses energy to power its muscles which push it into the air.

As a worm wriggles through the earth it uses energy to ripple its muscles which pull and push it through the soil.

Even the tiniest movement of your eyes as you read this page needs energy.

When an egg cooks it uses the energy from the cooker.

When an ice cube melts it takes energy from its surroundings.

Everything you or I do can be explained by energy too, that includes walking, talking and even turning this page. The energy to make you climb a hill comes from the energy in the food you eat.

Energy makes things happen. The more energy you've got the more you can do. The more you do the more energy you must give away. If you haven't got enough then you need to be given some in order to make something happen. You could say its a bit like money.

Energy supplies

The energy to make a car move comes from the chemicals in the petrol tank. The energy that causes so much damage and injury in a car crash comes from the movement energy of the car when it was travelling.

Whatever it is you need to do you need a supply of energy but it must be in the right form. For instance a car's energy comes from petrol, but petrol is poisonous to humans and cannot be used for energy.

Personal energy

Everything that we do needs energy to make it happen. To read this book takes energy, but not as much as you would need to run.

Even when you are asleep your body is using up energy. Your heart is pumping, you are breathing and your body is being kept warm. When you walk, run or do any exercise you are using up more and more energy.

So where does the energy come from? The food you eat keeps you alive. Food gives the energy to drive our bodies. If we stopped eating, our energy supply would be cut off and we could no longer stay alive.

How much energy do you need?

Energy is measured in joules or kilojoules (J or kJ).

If you are a small child you need 7 500 kJ each day.

A fifteen-year old girl needs 10 500 kJ a day

And a fifteen-year old boy needs 12 000 kJ a day.

A grown-up working all day at a desk would need less than the fifteen-year old boy - about 11 000 kJ a day.

Someone doing heavy work such as digging holes would need 16 000 kJ a day.

But how much food do we need to eat to get this energy?

A 100g apple contains 150 kJ of energy. At the other extreme a 100g chocolate bar contains 2,335 kJ. This meal would give you 6 338kJ of energy.

Bowl of tomato soup	(250 g)	703kJ
chicken	(200g)	1 544 kJ
potato chips	(250g)	2 473 kJ
peas	(150g)	410 kJ
ice cream	(150g)	1 208 kJ
Total		**6 338 kJ**

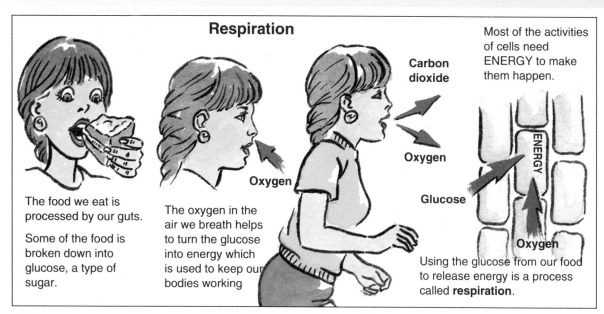

Respiration

The food we eat is processed by our guts.

Some of the food is broken down into glucose, a type of sugar.

The oxygen in the air we breath helps to turn the glucose into energy which is used to keep our bodies working

Carbon dioxide

Oxygen

Most of the activities of cells need ENERGY to make them happen.

Glucose

Oxygen

Using the glucose from our food to release energy is a process called **respiration**.

All living things respire, turning their food into energy for life. Some animals, such as mammals and birds use a lot of energy to keep warm. They are warm-blooded animals and have to keep their body temperature at about 37°C.

Cold-blooded animals have a temperature which may change when the surrounding temperature changes. Lizards bask in the sun to warm themselves up to their 'best' temperature for living. At night their temperature drops, because the sun's heat cannot warm them up.

Cold-blooded animals need less food than warm-blooded animals. A snake can go for a week without eating, and there is even an example of a snake that lived for over a year without eating!

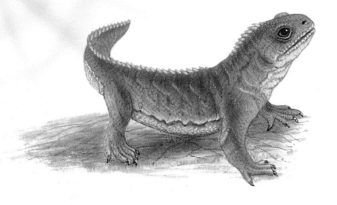

Plants and animals

Plants

Plants catch the sun's energy using a green pigment called chlorophyll. This traps light energy to make glucose from carbon dioxide and water. This is called photosynthesis and provides the plant with the energy it needs to live. Some plants change the glucose into starch which can be stored more easily than sugar.

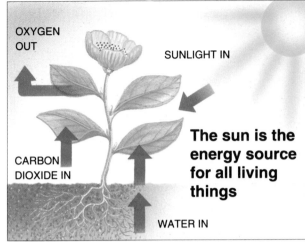

OXYGEN OUT

SUNLIGHT IN

CARBON DIOXIDE IN

The sun is the energy source for all living things

WATER IN

Plant storage systems include: bulbs, tubers and tap roots

All these parts of the plant and other parts like the leaves, stems, flowers and even the bark of trees are food for animals. Animals cannot make their own energy from the sun, so they take their energy from plants.

Bulbs: these are next year's leaves ready to grow when conditions are right, for example onions

Tubers: potatoes are the starch stores of the potato plant.

Tap root: Carrots are the swollen roots of the plant filled with stored starch.

Food Chains

Plants are the start of a food chain. Some animals eat plants for their energy. Other animals eat these animals for their energy. So energy from the sun is passed from plant to animal along this chain until it reaches the animal at the end of the chain. At every stage in the food chain some energy is lost so that there is less at the end of the chain than at the beginning. This means that there can only be a few animals at the end of the chain. This can be shown as a pyramid of numbers. It is possible to see this on the plains of East Africa just by counting the different animals in different parts of the food chain. There's lots of grass to provide a regular supply for herds of grass-eating animals such as gazelle. But there are fewer meat-eating animals such as lions, hyenas and leopards. There just would not be enough energy to go around to feed huge herds of meat-eaters.

Because energy is precious, animals do not waste it. Lions, tigers and other big cats spend much of their time resting. They may only eat once every three or four days and so use their energy carefully.

In winter many animals, such as dormice, have to hibernate because there is not enough food around to keep up with their energy demand . They drop their body temperature and slow all their life functions down until the spring. They use stores of fat built up during the summer to keep them going.

Storing Energy

When food is short as in the desert, animals find ways to store it.

The bodies of some honey-pot ants swell with the sweet sugary nectar which they collect from plants. Ants with the colony use these living larders hanging from the roof of the nest to feed the other ants in .

Humming-birds take nectar from flowers as their main source of energy. To make sure they do not waste their precious energy, their body temperature drops between each feed so less energy is wasted keeping themselves warm.

Energy and agriculture

Human beings have spread all over the world. How have they become so successful? Their big brains helped them to discover new ways of organising their most basic need - energy.

Hunter-gatherers

Some of the earliest human beings were hunter-gatherers. By hunting animals and gathering food such as nuts, fruits, leaves, seeds, eggs, roots, and even insects such as honey-pot ants, and grubs, they could survive in their environment. Some people still live like this today eg Kalahari Bushmen in parts of Africa. The numbers of hunter-gatherer people would have been quite small. The reason for this is that a great deal of land would have to be covered by these people to get enough energy to live.

The settlers

About 10,000 years ago people began settling in one place and not travelling around looking for food by hunting and gathering.

This happened because these people had found a way to grow their own plants, and to keep their own animals. They started to grow some plants in one place.

These were the first crops. They could breed goats and milk them or kill them when they wanted meat. They were no longer dependent on the wild environment. When the winter or a dry season came they could have enough to see them through these difficult times. They had found a way to control the amount of food they needed to live.

The first civilisations

It was not until people found ways to grow more food than they needed that civilisations came about. They began in places like Egypt along the river Nile and in Iraq along the rivers Euphrates and Indus. Each winter these huge rivers would flood into the surrounding land, fertilising it with mud and silt. By building special irrigation channels the people could control the flood waters and use them to enrich the soil for growing their crops. This natural fertiliser meant that they could grow more food than they needed and store it to eat later. Not everyone needed to spend time growing food. People could do other things. The pyramids of Egypt could only be built because of the 'spare' energy around to feed the architects and builders who created these magnificent structures.

Over thousands of years in different parts of the world whole civilisations came and went, first prospering and then collapsing. The Mayan civilisation in Central America was remarkable for its buildings, technology and advanced way of life, but in the end the numbers of people probably became too many fo the land to support and the civilisation collapsed.

People Increasing

As soon as fossil fuels were used to improve the way in which we grew our food the world food supply increased and so did the number of people.

The world's population is enormous compared to two thousand years ago, and it is still increasing.

The Industrial Revolution

The Industrial Revolution

The Industrial Revolution changed the world and it was the change from one fuel to another that helped to make it happen. Before the Industrial Revolution wood was the most important fuel in Britain. In the reign of Queen Elizabeth I (four hundred years ago) coal was used by brewers, but it was not popular because of the smoke and fumes. Firewood was becoming difficult to obtain and was expensive. In the 1600s, there was not enough land to grow crops or keep animals to feed the growing population. Land was also needed to grow trees which were used for building ships and homes as well as for firewood and charcoal. Land was also needed to grow grass and hay to feed the vast numbers of horses, which were the only transport. Wool was a large part of Britain's trade with other countries and land was needed too to raise sheep. Land was running out, and a new, cheaper fuel was needed.

From wood to coal

At the time of the Industrial Revolution firewood had become very expensive and so the industries of the time, such as brewing, making soap and candles began to use coal. Coal became a very important fuel once new ways of making iron were discovered.

Coal was originally mined by digging at the surface of the earth. But to find more, coal miners had to dig deeper into the ground. But as they dug deeper they found water and many mines flooded. Thomas Newcomen invented the first steam engine in 1712 to pump water out of the mines.

The first steam engines

In 1763 James Watt improved the steam engine so that it used up less energy. This engine was used to drive the machines for industry, such as making cotton. New towns and cities sprung up all over the country as industries grew. Most people had worked in the countryside, now they came to new jobs in factories in the towns living and working in crowded, and unpleasant conditions. Each industry needed coal and raw materials.

Nineteenth century slums in London

Early transport

The next energy revolution came in the way heavy materials were transported around the country. First canals were built. These could carry large supplies of heavy materials needed by industries, but they were slow. They also needed horses to pull barges along them. It was the invention of steam trains and the railways which really improved the transport of goods and people.

The first locomotives

The first train was built by Richard Trevithick for the Penny Daren iron works and it pulled ten tons of iron and seventy men 15 kilometres. But it was George Stephenson's 'Rocket' engine invented in 1829 which made a real success of this new form of transport. The railways became the most important form of transport for raw materials and manufactured goods.

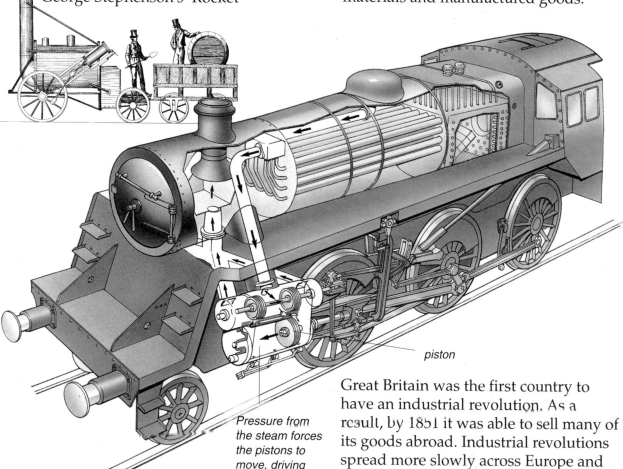

piston

Pressure from the steam forces the pistons to move, driving the locomotive

Great Britain was the first country to have an industrial revolution. As a result, by 1851 it was able to sell many of its goods abroad. Industrial revolutions spread more slowly across Europe and the world.

Solid fuels

Solid Fuels

All living things, animals and plants contain an element called carbon. Carbon is in the gas called carbon dioxide which is in the air and in the food we eat. The carbon in living things is borrowed from the Earth until the living thing dies. Through natural rotting or decaying the carbon is returned to the air and can become part of new living things. This is called the carbon cycle. Carbon locked-away in dead animals and plants provides fuel for use by people.

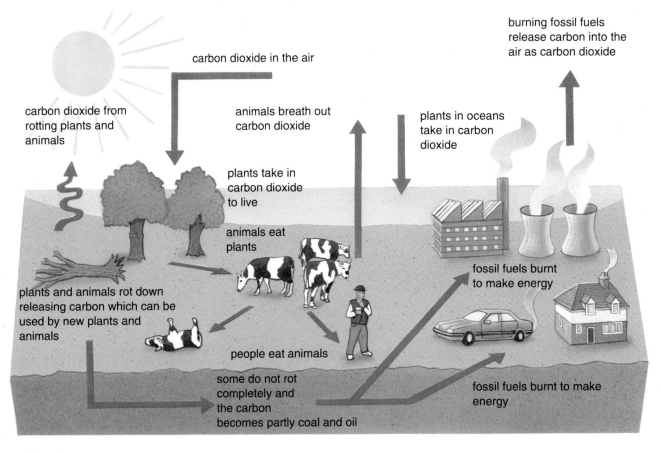

burning fossil fuels release carbon into the air as carbon dioxide

carbon dioxide in the air

carbon dioxide from rotting plants and animals

animals breath out carbon dioxide

plants in oceans take in carbon dioxide

plants take in carbon dioxide to live

animals eat plants

fossil fuels burnt to make energy

plants and animals rot down releasing carbon which can be used by new plants and animals

people eat animals

some do not rot completely and the carbon becomes partly coal and oil

fossil fuels burnt to make energy

Peat

In some places under the ground there is not enough oxygen for living things. The bacteria which normally rot things cannot survive in these places. So when plants die they are not completely rotted and make peat. Peat is an early stage of coal formation. In peat there is a great deal of carbon which can be released into the air by burning. At the same time heat is released, so people dig up peat and use it as a fuel.

Peat moss drying in Ireland

Coal

Coal is formed in a similar way to peat. Most coal comes from plants that lived about 300 million years ago at the time known as the Carboniferous era. The plants around then were giant ferns and horsetails. When they died they did not rot completely. Eventually sand and mud covered them. After millions of years they were covered by deep rocks. The pressure of the deep rocks and heat from the Earth turned the plants into coal.

Carboniferous plants...

...rot down...

...turn into coal...

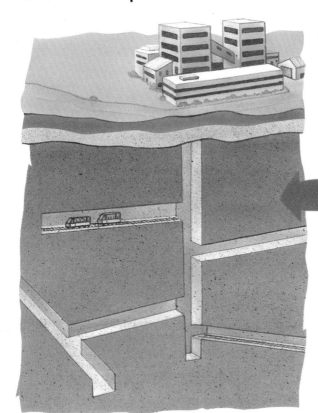

...which we mine

The coal we burn today is made from the carbon that was trapped in the plants 300 millions years ago. The more coal we burn the more carbon we release into the atmosphere, adding to global warming. We are using this valuable fuel up very quickly.

Burning

Burning is a good way of releasing energy for light and heat.

The heat from this fire comes from chemical energy that was stored in the wood.

The light from this candle comes from the chemical energy stored in the wax

Good fuels and burning

A good fuel is one which releases a lot of energy when it burns. Fuels which burn well are usually made up of the two chemicals hydrogen and carbon. The energy is stored in the links between these two chemicals. Energy is released when these joins are broken and the chemicals join up with other chemicals such as oxygen. Links between hydrogen and oxygen and carbon and oxygen need less energy than links between hydrogen and carbon. It is easier for hydrogen and carbon to join up with oxygen and this is what happens when a fuel burns. The extra energy that is not needed to hold the hydrogen and carbon together is released as heat and light.

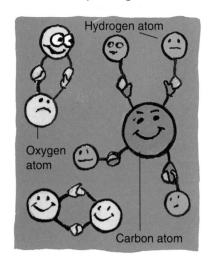

1 Before
Atoms joined together to make oxygen and fuel.

2 Heating
Heating causes the joins to break up.

3 After
Atoms join up again but in a different way which needs less energy, this releases energy.

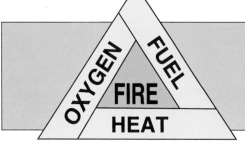

To get a good fire you need fuel and oxygen, but a fuel will not burn unless it is at the right temperature. So it needs heat to get it going, that's why you need a match to light a candle or a spark to light the gas. Once the fire has started then it releases energy which heats the gas and keeps the flames burning.

Fires can get out of hand. Nearly everything around us contains hydrogen and carbon will burn. A small spark can start a fire and release energy. As the fire heats up the surroundings they too can burn and can release more energy. There are 3 ways of stopping this:

1 Cool the fuel down so it cannot get hot enough to burn. This is what these fire hoses are doing. ▶

◀ **2** Starve the fire of oxygen. This is how this fire blanket works.

3 Make sure that the fire ▶ runs out of fuel. This is how the Great Fire of London was stopped, houses were knocked down so that the fire had no more fuel to burn.

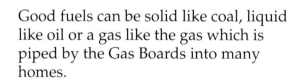

Good fuels can be solid like coal, liquid like oil or a gas like the gas which is piped by the Gas Boards into many homes.

The gas in a gas cooker is a good fuel because it mixes easily with air which contains oxygen needed for burning. Burning also produces harmful waste gases which must be able to escape. Modern gas fires and boilers have special pipes called flues which take the waste gases and fumes away.

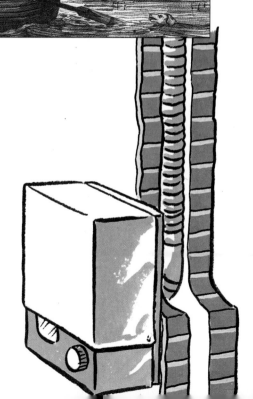

Oil and gas

How oil is formed

Minerals oils take many billions of years to form underground. They are being formed today whenever tiny organisms die and submerge in mud. Normally most tiny organisms rot in the earth's atmosphere but some get squashed and heated in the same sort of way that coal is formed. This squashing and heating causes chemical changes which are helped by special bacteria. Eventually the carbon and hydrogen join to form oil. This slowly becomes liquid and moves through tiny holes and cracks in the rocks until it reaches the surface or is trapped under rocks which will not let it through.

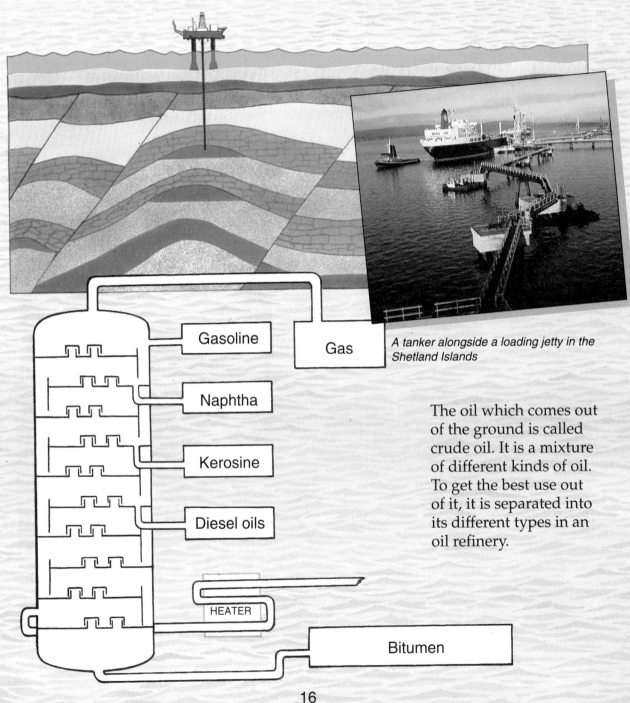

A tanker alongside a loading jetty in the Shetland Islands

Gasoline

Gas

Naphtha

Kerosine

Diesel oils

HEATER

Bitumen

The oil which comes out of the ground is called crude oil. It is a mixture of different kinds of oil. To get the best use out of it, it is separated into its different types in an oil refinery.

Natural gas

A gas called methane is formed in the same way as oil. Deep in the rocks of the earth carbon and hydrogen can be found together, sometimes at the top of an oil trap. Once this was just a nuisance and was burnt off in a spectacular flame just to get rid of it. Now it is recognised as a valuable energy source. This natural gas is piped to many homes throughout the United Kingdom and Scandinavia from rocks under the North Sea. To get to our homes it is pumped along pipes over the seabed to the shore. Here it is treated before being fed into the national network of pipes. Little treatment is needed but it has to be checked and purified, this process is called scrubbing.

Natural gas has been used as a fuel in the USA since the discovery of gas fields in Texas in 1918.

natural gas so that it was almost like coal gas or to change all gas appliances so that they could run on natural gas. It was decided that although this was a major job it would make the best use of the energy supply if the whole system was changed to run on the new natural gas. All the old coal gas left in the pipes was burnt and every gas appliance in every home was modified.

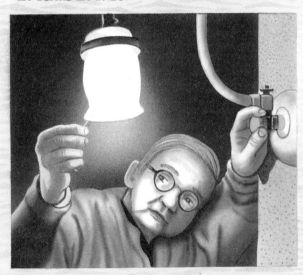

Natural gas was not discovered in the North Sea until 1959. Up till then UK gas was supplied locally from gas works and was called Town gas. It was produced from coal and was also known as Coal gas. This provided the first form of street lighting.

Natural gas offered a plentiful supply of energy but was slightly different from coal gas. An important decision had to be made. Either to try and treat the

Some homes use gas but are not close enough to a piped supply or gas main. Houses like this one keep a canister of gas under high pressure. This one has been surrounded by shrubs to make it more sightly. The gas has been squashed under pressure until it becomes liquid. It takes up less room and is easier to transport and store. The canister has to be very strong because the gas is under high pressure.

Research has meant that modern fires and boilers are much more efficient than older appliances. It is important that research continues to make sure we waste as little energy as possible.

Energy in the home

A modern home uses energy in many ways

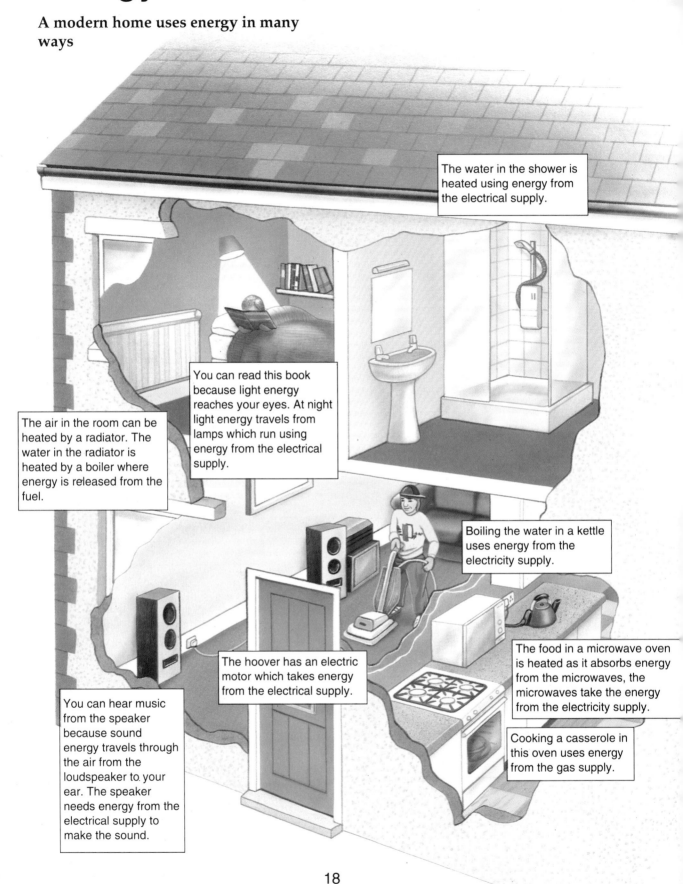

The water in the shower is heated using energy from the electrical supply.

You can read this book because light energy reaches your eyes. At night light energy travels from lamps which run using energy from the electrical supply.

The air in the room can be heated by a radiator. The water in the radiator is heated by a boiler where energy is released from the fuel.

Boiling the water in a kettle uses energy from the electricity supply.

The hoover has an electric motor which takes energy from the electrical supply.

The food in a microwave oven is heated as it absorbs energy from the microwaves, the microwaves take the energy from the electricity supply.

You can hear music from the speaker because sound energy travels through the air from the loudspeaker to your ear. The speaker needs energy from the electrical supply to make the sound.

Cooking a casserole in this oven uses energy from the gas supply.

How much energy?

Some things use energy very quickly others more slowly. A kettle will use electrical energy quickly heating the water for a bath. A light bulb will use less energy than an electric fire if they are switched on for the same length of time.

If we measure energy by what you can do with it, then one way is to compare the amount of energy used with the amount needed to boil a kettle full of water.

	To heat a bath full of water would take the same amount of energy as boiling about twenty kettles. x5 x5 x5 x5
	To heat the average house for one evening in winter would take the same amount of energy as boiling about two hundred and forty kettles. x240
	To light a room for one evening would take the same amount of energy as boiling about three kettles.
	Watching two hours of television would take the same amount of energy as boiling seven kettles.
	A washing machine uses the same amount of energy as boiling about ten kettles.

Paying for it

Energy for our homes has to come from somewhere and must be paid for. Most homes have an electricity meter which measures the amount of electrical energy used. So the more kettles you boil the more you have to pay

It is possible to measure energy in kettles, but it is more usual to measure energy use in kilowatt hours. One kilowatt hour of electrical energy is enough to boil about five kettles.

Getting energy into the home

Homes are built to take account of the way energy gets to the home.

Homes built in the 1920s often had a coal cellar for storing fuel. These were often built so that they could be loaded easily from the road. Coal fires were used to heat most homes so chimneys were a common part of the landscape.

Today many people have their energy piped to their homes. If a house uses gas and electricity both are usually supplied from outside the home. There will be separate meters for electricity and gas.

Using electricity

An electric current flows in wires. We use the energy supplied by electricity for four main purposes.

For running electric motors such as in a hair dryer, washing machine and fridge.

For lighting

For heating

For communication

All these use energy. We also use a tiny amount of electricity to control these, by working the central heating system for example.

The circuit is the pathway the current must follow, so there must be no breaks in it. Usually the circuit is made of metal wires. Metals and other materials which allow electricity to pass through them easily are called conductors. Electricity can hardly travel through plastic at all. The metal wires are often covered in plastic to make sure the electric current keeps to the right path. Materials like plastic which do not allow electricity to pass through them easily are called insulators.

An electric circuit

When the paperclip is moved to toach the drawing pin the circuit is complete and the bulb glows

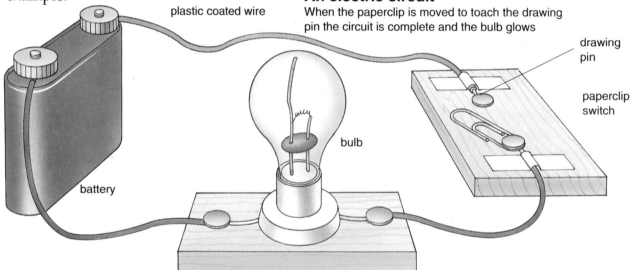

plastic coated wire

drawing pin

paperclip switch

bulb

battery

Where does electricity come from?

Thunder and lightning release short bursts of electrical energy. You can get similar effects when you take off a nylon shirt. If you do this in the dark you may see the sparks as well as hear a crackle. This would not be a reliable source of energy Batteries are a more reliable energy store because they can give a steady supply of electricity and are easy to carry about.

Batteries

The first battery was made by Alessandro Volta about 200 years ago. It was made of two different metals such as copper and zinc in salty water. The electricity was made by the chemical reaction between the copper and zinc in the salt water. Modern batteries work in the same way using specially developed chemicals.

Volta's column pile, 1800

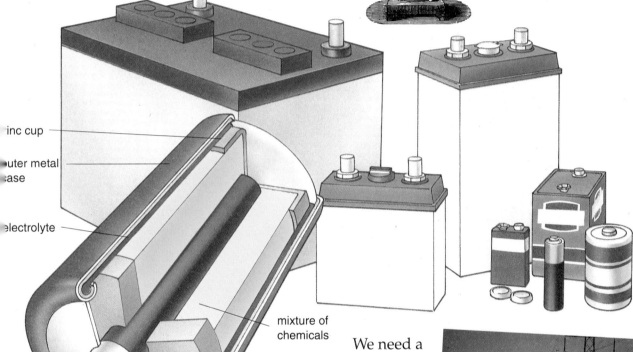

inc cup

uter metal
ase

electrolyte

metal cap

Cutaway of a modern dry-cell battery

mixture of chemicals

The zinc reacts with the chemical contents and electrolyte to produce an electric current

A battery is a store of chemical energy which can be released as electrical energy. Batteries run down when the chemical reaction is complete. Some batteries can be recharged using electrical energy. Some chemicals in batteries are poisonous so you need to throw away old batteries carefully.

We need a reliable supply of energy in our homes. If the electrical energy needed is low it can be provided by a battery as with a portable radio. But for machines like microwave ovens and washing machines which need a lot of electrical energy we use electricity from the national grid which is cheaper than using batteries.

Generating electricity

If you have a dynamo on your bicycle then you are generating electricity.

When the spindle is against the turning wheel then the magnet spins and an electrical current flows in the circuit making the lights come on.

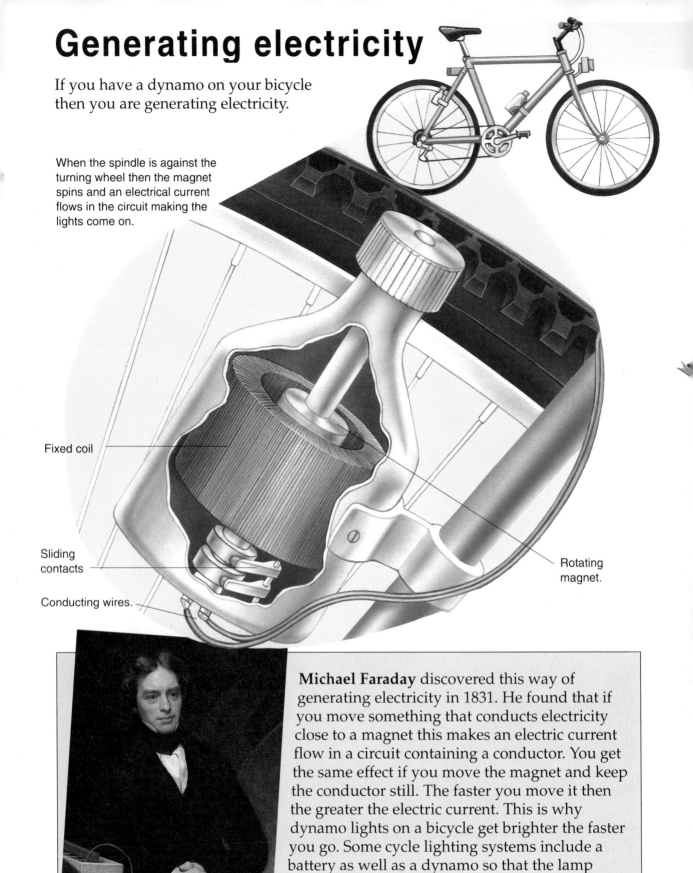

Fixed coil

Sliding contacts

Conducting wires.

Rotating magnet.

Michael Faraday discovered this way of generating electricity in 1831. He found that if you move something that conducts electricity close to a magnet this makes an electric current flow in a circuit containing a conductor. You get the same effect if you move the magnet and keep the conductor still. The faster you move it then the greater the electric current. This is why dynamo lights on a bicycle get brighter the faster you go. Some cycle lighting systems include a battery as well as a dynamo so that the lamp glows after you have stopped.

A power station makes electricity in a similar way as a cycle dynamo only on a larger scale. In a cycle dynamo you supply the movement energy when you pedal. In a power station the energy that produces the movement comes from the fuel.

Coal is one kind of fuel.

Coal-fired power station needs to be built where there will be a good supply of coal.

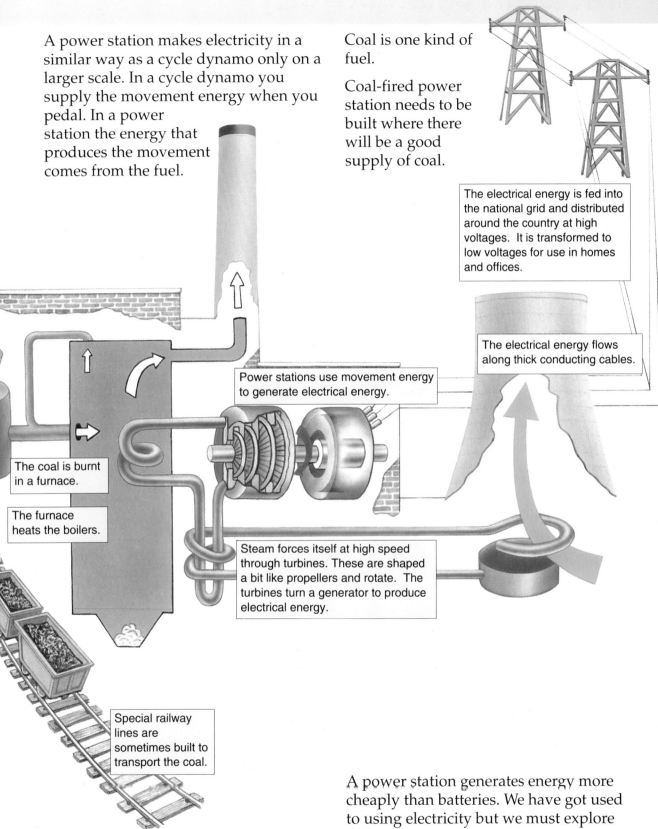

The electrical energy is fed into the national grid and distributed around the country at high voltages. It is transformed to low voltages for use in homes and offices.

The electrical energy flows along thick conducting cables.

Power stations use movement energy to generate electrical energy.

The coal is burnt in a furnace.

The furnace heats the boilers.

Steam forces itself at high speed through turbines. These are shaped a bit like propellers and rotate. The turbines turn a generator to produce electrical energy.

Special railway lines are sometimes built to transport the coal.

You don't have to use coal, you can use almost any energy source.

A power station generates energy more cheaply than batteries. We have got used to using electricity but we must explore different ways of providing energy and make sure we waste less.

Nuclear fuels

Atoms

Everything is made up of atoms joined together in different ways. Every substance has a different kind of atom. Once it was thought that the atom was the tiniest things there could be. But in 1910 scientists found even smaller things inside the atom at the centre. This clump of tiny particles at the centre of the atom is called the nucleus. Nuclear energy is released from the nucleus.

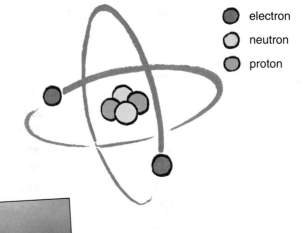

electron
neutron
proton

Nuclear fission

A nuclear power station is one which uses nuclear fuel. A nuclear fuel is one which releases energy from inside the atoms themselves. This happens when an atom is split up into smaller pieces.

For most atoms this is difficult to do. But some atoms, like uranium split more easily an release energy. The energy is released by uranium is made into fuel rods and can be used to generate electricity in specially designed power stations.

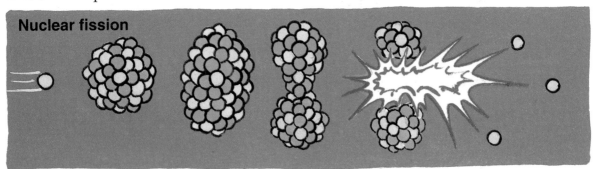

Nuclear fission

Waste

Eventually the Uranium 235 fuel rods lose their effect and cannot be used to generate electricity. Some of the old fuel can be turned into new fuel rods. Uranium fuel produces waste which contains radioactivity which must be carefully managed to protect the environment.

Uranium is a metal found in the rocks of the earth. It is mined from rocks in Australia, Canada, France and the USA and in some parts of the Commonwealth of Independent States.

Nuclear reactions and radioactivity

Some atoms are unstable, these are radioactive. To become stable they lose energy. This energy is released in two main ways, either by shooting off tiny particles at high speed or by sending out special waves of energy called ionising radiation.

Nuclear reactors produce energy by a process called nuclear fission. Uranium absorbs particles called neutrons into the nucleus of the atom. This makes the atom split and releases energy. This energy is used to power an electricity generator.

If radiation escapes from a nuclear reactor it can be dangerous. High speed particles may damage the structure of the body's cells. Fortunately it is easy to protect us from this energy with the right shielding but it is important not to touch radioactive material because radioactive particles could get inside you where they will cause much more damage.

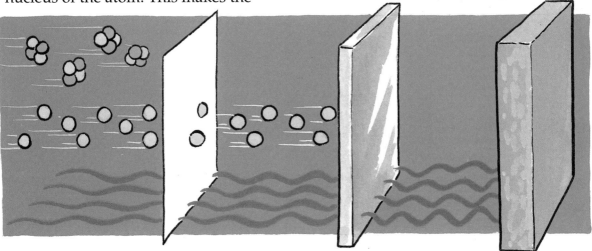

Nuclear fusion

Some atoms can be squashed together to make another atom. The new atom needs less energy than the original two atoms, so the extra energy is released. Fusion reactions happen in the sun all the time to make heat and light energy.

Scientists are trying to produce the same kind of nuclear reaction here on earth so that they can be used to generate electricity in power stations. The main fuel is hydrogen , so if scientists are successful then there is plenty of fuel using the hydrogen in sea water.

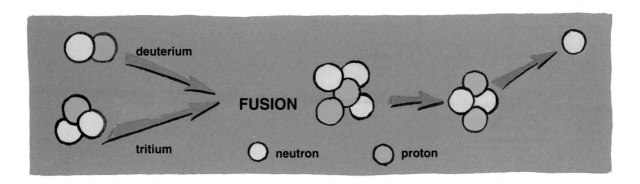

deuterium

tritium

FUSION

neutron

proton

Renewable energy

Coal, oil and gas are all fossil fuels. They are found in the earth and were formed millions of years ago. Once all the coal is dug up, all the oil and all the gas is pumped out of the ground there will be none left. Fossil fuels are called non-renewable energy sources, because once used they can never be renewed.

Some kinds of energy will not run out and are known as renewable energy sources. Renewable energy comes from the sun, from wind, from the oceans' waves and tides, from rivers, from the heat of the rocks deep in the Earth and from plants. Because renewables are part of the Earth's natural cycles they will always be renewed.

People have been using these sources of energy for a long time and many still rely on them all over the world.

The earliest explorers relied on the wind blowing into a sail to drive them across the sea.

The windmill uses the wind's energy to drive the grindstones, to grind wheat to be turned into flour to make bread. Smaller windmills are still used today to pump water up from wells.

The power of water flowing downstream could drive huge waterwheels which drive millstones for grinding cereals such as wheat and corn.

Early civilisations in hot countries built thick walled houses. These took up the sun's heat during the day keeping the rooms inside cool. But at night the stored heat passed into the cool rooms warming them up.

For centuries people have burnt wood as fuel. In many countries wood is still the most important source of fuel for cooking. In parts of Europe the woodlands were coppiced. This means that the trees are cut just above the ground. Hardwood trees will sprout many new shoots even when they are cut. After four or five years the shoots are cut off and used for firewood. The trunk of the tree will then make more shoots, which can be harvested again. This is quicker than cutting down the whole tree and waiting for it to grow.

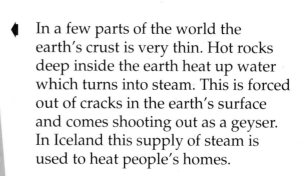

In a few parts of the world the earth's crust is very thin. Hot rocks deep inside the earth heat up water which turns into steam. This is forced out of cracks in the earth's surface and comes shooting out as a geyser. In Iceland this supply of steam is used to heat people's homes.

In some parts of the world where there is a shortage of trees, people collect animal dung and dry it in the sun. It is then used as a fuel and burnt to make fires for heating or cooking.

A geyser in Iceland

More about renewable energy

Although there is a great deal of energy available from natural, renewable resources it is not always found where it is needed. Relying on wind energy makes sense in windy places - most windmills are found on open land where the wind blows without anything to stop it - but could be the wrong kind of energy supply in a sheltered valley.

Renewable energy is difficult to control and needs modern technology to make use of it. At the moment fossil fuels are much cheaper to burn so people keep using coal, oil and gas rather than renewable energy sources.

Different technology is used to make the most of this endless natural energy.

A wind mill for pumping up underground water in Kenya, Africa

Solar power

◄ The sun's heat is collected using a special flat plate collector. The plate is usually black, and absorbs the heat of the sun. Pipes carrying water pass through the plate and the water heats up. This water is then pumped to where it is needed. In some countries, such as Israel most of the water in people's homes is heated by solar power. Solar cells turn the energy in sunlight straight into electricity which can be used to power anything from a calculator to a satellite.

A solar power station at Odeilo in France

Wind power ►

Today there are some places in the world which are windy enough to use modern windmills to produce electricity from the wind. This wind farm in California can produce several hundred kilowatts of power. Although a clean source of power, wind farms can spoil the appearance of the natural environment and can be very noisy too.

Wind turbines at Palm Springs, USA

Water power

Energy from water is often used in places with mountains and rivers. Countries such as Canada, Scotland and Norway use energy from falling water, hydro-electric power, to drive turbines which make electricity. Large dams are often built to supply water and hydro-electric power.

Power from the ocean

Turbines like the ones used to make hydro-electric power can be used to turn the rise and fall of the oceans' tides into electricity. These are usually found in estuaries where the mouth of a river empties into the sea.

Energy in the oceans' waves can also be used to generate electricity.

Bio-fuels

Bio-fuels come from dead plants or rotting material.

In Austria farmers grow oil-seed rape which they turn into Bio-diesel for their motor vehicles. It takes 1/8th of an acre of oil seed rape to produce 14 gallons of Bio-diesel, which is 850 miles of motoring.

The fuel is much cleaner than ordinary petrol and diesel. Because the carbon in the plant came from the atmosphere in the first place, burning the fuel does not add extra carbon to the atmosphere and so does not add to global warming.

Oil seed rape in Lancashire, England

Anything that rots produces methane and some gas can be piped off from huge rubbish dumps to be used just like any other natural gas for heating homes.

Making things hot

How do things change
as they get hotter?

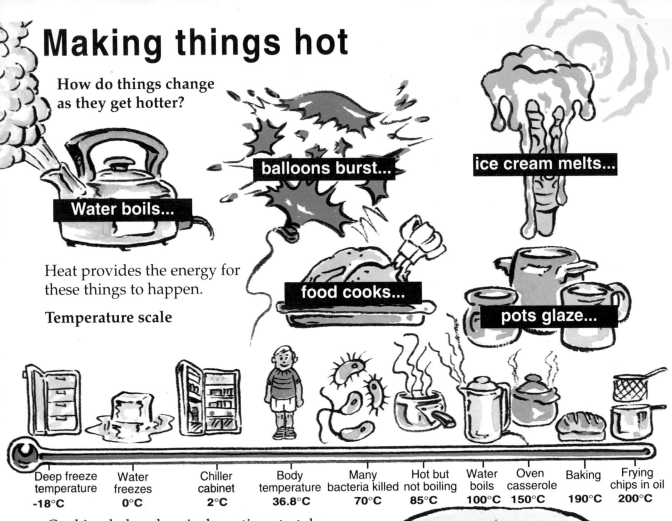

Water boils...

balloons burst...

ice cream melts...

food cooks...

pots glaze...

Heat provides the energy for
these things to happen.

Temperature scale

Deep freeze temperature	Water freezes	Chiller cabinet	Body temperature	Many bacteria killed	Hot but not boiling	Water boils	Oven casserole	Baking	Frying chips in oil
-18°C	0°C	2°C	36.8°C	70°C	85°C	100°C	150°C	190°C	200°C

Cooking helps chemical reactions to take
place in food which change its flavour.
Cooking also kills germs. Heating food
thoroughly to a temperature greater than
70° will kill most most germs usually
found in food. although it may not
destroy the poisons produced by them. If
food is not heated thoroughly, bacteria in
the centre may not be heated enough to
kill them. They may only be warmed
and use the heat energy to divide and
grow. At low temperatures, bacteria do
not have enough energy to multiply, so
many foods are kept in the fridge before
cooking.

Different foods take different times to
cook. Food in an oven is heated from the
outside, so the time taken for the heat to
pass right to the centre depends on the
food. Baked potatoes cook slowly in an
oven but tomatoes cook quickly.

In European homes we use gas or
electricity to provide heat energy for
cooking every day. In the summer
barbecues use charcoal to provide the
heat.

Microwaves, like radio waves or the waves from the sun carry energy. Sometimes they can travel through things, sometimes bounce off and sometimes they are absorbed and their energy heats up the material that absorbs them. For instance, microwaves pass easily through paper but not through water. Instead they are absorbed and their energy heats the water. Water is one of the main ingredients of the food we eat so that microwaves can be used for cooking. Microwaves travel through the outer layers of food until they are absorbed by water inside the food. This means that the inside heats up first and not the outside in as in an oven. This means that is is difficult to cook things so that they have a crispy surface.

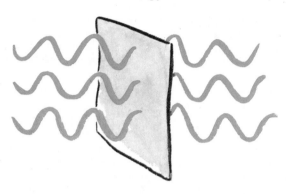

Microwaves pass easily through paper

Microwaves are absorbed by water which is heated by their energy

Pottery hardens in a kiln because the heat energy causes chemical reactions in the clay which hold the particles more firmly together. Glazes which gives pots shiny surfaces melt at high temperatures but become solid on cooling. Notice how the kiln is built with vents to allow air to feed this flames.

Blast furnaces need enormous amounts of energy to separate the iron from the rocks which contain the iron ore. ▼

Insulated metal bucket

A charcoal stove
Clay lining helps to keep the heat in so that less fuel is needed

Air inlet

Keeping things warm

Energy and heat

When something heats up it is absorbing energy, for example your skin feels warm when you absorb energy from the sun.

If a hot meal is left in a cold room then the food loses energy to the cold room and cools down.

To keep things warm we need to slow down or stop heat energy from escaping from hot things to their cooler surroundings. One way to keep a hot meal warm would be to put a cover on it. What other ways have people and animals found to stop heat energy escaping?

How energy travels

Energy can travel from hot areas to cold areas in three ways:

1 By being passed from one particle to the next through a solid. This is why the handle of a teaspoon becomes hot when you stir a hot drink. This movement of energy is called conduction.

2 By heating up the air around it and mixing it with the surrounding cold air. This causes air currents which replace the warm air next to the hot object with more cold air. This is the way a hot meal left on the table becomes cold and is called convection.

3 By radiating heat energy as waves which can travel through empty space. The hotter something is the more energy it releases. This is how energy reaches us from the sun and the heat you feel from an electric fire reaches you in the same way. This is called radiation.

Polar bears have a thick layer of fat and thick water-proof fur to stop their heat energy escaping too quickly.

Birds cannot stop heat energy escaping completely, so to keep its body warm birds need to replace the energy that is lost. To do this they use energy from the food they eat. That's why it's important for birds to find food in cold weather.

Humans do not have fur, only a thin layer of body hair. Clothes which trap air next to our skin are best in winter for keeping us warm unless they become damp.

In winter blackbirds ruffle their feathers to trap air, this too makes it more difficult for energy to escape.

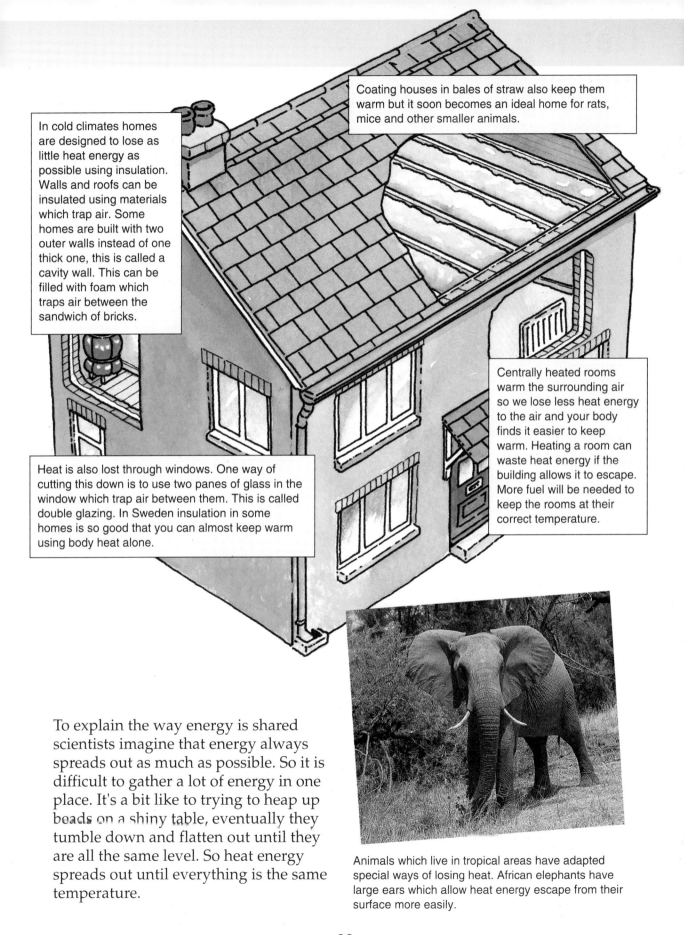

In cold climates homes are designed to lose as little heat energy as possible using insulation. Walls and roofs can be insulated using materials which trap air. Some homes are built with two outer walls instead of one thick one, this is called a cavity wall. This can be filled with foam which traps air between the sandwich of bricks.

Coating houses in bales of straw also keep them warm but it soon becomes an ideal home for rats, mice and other smaller animals.

Centrally heated rooms warm the surrounding air so we lose less heat energy to the air and your body finds it easier to keep warm. Heating a room can waste heat energy if the building allows it to escape. More fuel will be needed to keep the rooms at their correct temperature.

Heat is also lost through windows. One way of cutting this down is to use two panes of glass in the window which trap air between them. This is called double glazing. In Sweden insulation in some homes is so good that you can almost keep warm using body heat alone.

To explain the way energy is shared scientists imagine that energy always spreads out as much as possible. So it is difficult to gather a lot of energy in one place. It's a bit like to trying to heap up beads on a shiny table, eventually they tumble down and flatten out until they are all the same level. So heat energy spreads out until everything is the same temperature.

Animals which live in tropical areas have adapted special ways of losing heat. African elephants have large ears which allow heat energy escape from their surface more easily.

Getting about

The Vikings explorers used the energy in the wind and tides to power their sailing ships. Or they rowed using a crew of slaves, this would have been hard work.

Modern ships are driven by powerful engines. Their energy comes from fuel which is a store of chemical energy. Ships usually use diesel oil as fuel.

The engine works by making the fuel explode releasing chemical energy which pushes a piston out with terrific force.

The piston movement turns the propeller. This turns like a screw and pulls the ship through the water.

Getting Going

Some things need to move fast using a lot of energy in a single burst. A grasshopper must move quickly to dodge its enemies. It has powerful leg muscles for jumping suddenly, these use energy.

When an arrow is fired from a bow it is given a sudden burst of motion energy.

The bow is relaxed

The archer pulls back the string using energy.

The string and bow are now tight. It is ready to fire the arrow, it has energy.

When the string is released the stored energy is released and the arrow is pushed forwards.

The bow has lost its energy and the arrow has motion energy.

Going far

When an arrow is fired or a ball is thrown, energy for movement is supplied once, at the very beginning. If you want to go a very long distance then you need to be able to take your energy supply with you.

If you are going to take part in the Tour De France cycle race then your training will include making sure you eat the right kind of food to give you an energy store for the long ride.

A petrol tank carries the energy supply for a car.

An electric train uses electrical energy supplied along its route in overhead wires or a special conducting rail which runs alongside the rail. Never play near a railway line.

Friction and keeping going

If you slide a book along a table it will eventually stop because of a force called friction. It is as if this force pushes back against an object when it moves across a surface.

A sled moves well on ice but not on a dry path. You need to keep pushing it to keep it going. Each push gives it movement energy to replace the energy lost in friction between the path and the sled runners. This lost energy makes the sledge runners hot.

Wheels move across a rough surface better than a sledge because they waste less energy. A trolley pushed along a path will eventually stop because the moving wheels waste energy. This wasted energy warms up the wheels instead of keeping them moving. If moving parts are oiled they slide over each other more easily and less energy is wasted as heat.

At slow speeds sliding things over water is better than using wheels because there is less friction. This is one reason why canals were used to transport heavy goods.

Sliding on air wastes even less energy because there is less friction. A Hovercraft slides on its own cushion of air underneath and not on the ground or water surface.

Force

If you want to jump, throw a ball or fire an arrow you need energy from your muscles to push or pull. This push or pull is known as a force.

35

Going fast

To go fast you need to use a lot of energy which is why running fast makes you tired. Have you noticed that when you run fast you need to breathe quickly? That's so that your body can get enough oxygen to release the energy stored in your muscles.

A shark can swim quickly through the water because it has powerful muscles in its tail and is a good shape for moving through the water, it is streamlined.

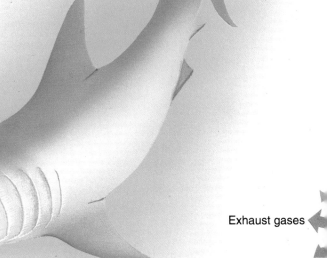

Exhaust gases

Going faster

The faster you a bike goes the greater the friction between the moving parts and between the wheels and the ground. When you travel fast there also is friction between the bike and the air itself which wastes valuable energy. That's why racing cyclists sometimes wear specially shaped helmets to cut down friction between themselves and the air. These helmets are streamlined. Cars are designed to be streamlined so that they waste less energy and use less petrol.

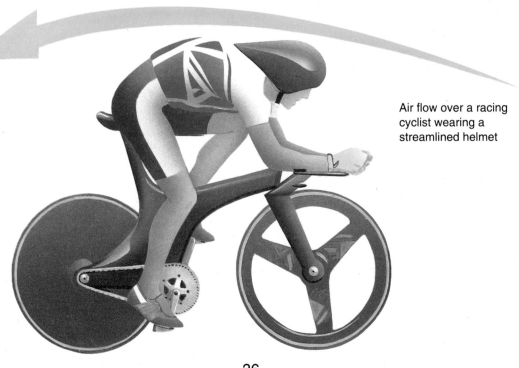

Air flow over a racing cyclist wearing a streamlined helmet

A modern aircraft will use a jet engine instead of an ordinary piston engine. This uses energy quicker and without waste.

A jet engine throws out hot gases from the back of the jet. The faster the gases are thrown out, the faster the aircraft travels forward.

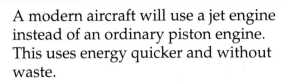

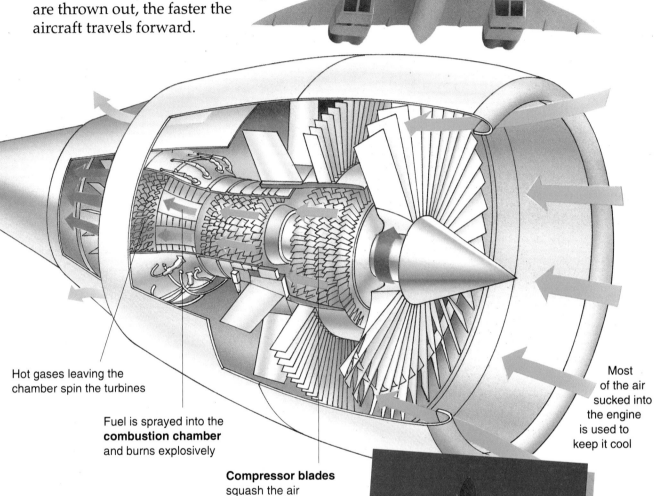

Hot gases leaving the chamber spin the turbines

Fuel is sprayed into the **combustion chamber** and burns explosively

Compressor blades squash the air

Most of the air sucked into the engine is used to keep it cool

A rocket launcher works in the same way. This uses a lot of fuel very quickly in order to climb away from the earth and push its way through the air. Once it leaves the earth's atmosphere there is no air to slow it down, only the pull of the earth In fact in deep space if the engine were switched off then the rocket would carry on travelling for ever without using any fuel until it hit something.

Modern agriculture

Fossil fuel gave the energy to drive the Industrial Revolutions. These same fuels have been used to change the way that food has been produced all over the world. The more fossil fuels that are put into growing food, the more food that can be grown and the greater the number of people that can be fed. So the more food that is produced, the cheaper it is to buy.

Traditonal farming in Northern Thailand

Before fossil fuels most people would have been involved with the growing of food.

People ploughed the land with ploughs pulled by cattle. They sowed the seeds by hand and harvested the crops with a scythe. They milked the cows by hand. These methods were still used as recently as 25 years ago in parts of Europe. Some people today in different parts of the world still rely on their own muscle power and the help of animal muscle power.

Today, modern technology and fossil fuels have replaced the human energy that was needed to grow food in many parts of the world. (In fact the growing of food has become so successful for some that it is no longer worthwhile for farmers to produce food on all of their land.)

Extra energy is needed to drive the farm machinery to do the jobs that people used to do. (For example tractors pulling ploughs and seeding machines, combine harvesters harvesting the crops.)

But there are other areas of modern agriculture that have been affected by the extra energy available.

Fertilisers are needed to top up the chemicals taken out of the soil by the crops. Huge factories make these fertilisers and they are spread by the tonne over the fields. When the crops grow they are attacked by different pests so chemicals called pesticides are sprayed over the fields to keep the pest numbers down. Weeds try to grow between the crop plants so crops are sprayed with herbicides to stop the weeds growing. Sometimes they need several different sprayings of chemicals to make sure they grow well.

All this involves energy which is provided by fossil fuels which is why people are able to produce extra food in some parts of the world. In fact the growing of food has become so successful that some can be stored to be eaten later.

There are problems with modern agriculture. The soil is used so much that the only way that farmers can keep growing crops is by adding fertilisers and pesticides. But the earth's soil is disappearing at an alarming rate, blown away by winds or washed away by rain.

Also the pesticides used to kill insects can poison the soil and animals which eat the poisoned pests can be made sick by the pesticides which collect in their bodies.

CASE STUDY

Wall painting from the tomb of Sennejem showing farming and irrigation canals along the River Nile.

For thousands of years the River Nile in Egypt has flooded the surrounding flat lands bringing mud and silt which naturally fertilises the soil. When the Aswan Dam was built across the River Nile, it stopped most of the silt from reaching the land. Farmers had to buy artificial fertilisers to put on the soil. These were made in factories whose energy came from the dam's hydro-electric turbines. So although the dam created one energy problem it helped to solve another. The use of this fossil fuels to provide extra energy has meant that there is more cheap food. But there are problems with fossil fuel pollution which may affect peoples' health. Other cleaner fuels would be expensive to use which would put up food prices.

The environmental impact of energy

When the first human beings discovered fire and how to keep it burning they did not know that their descendants would carry on making fire which would put the natural environment at risk.

Fire kept early people warm, it could also turn cold raw food into hot food. Later it was used to turn the mineral ore from the rocks into tools and weapons. Fires have been used ever since which has caused rapid destruction of the environment.

Destruction of forests in Sumatra, Indonesia

Two thousand years ago much of Europe was covered in thick forest. Nearly all of it was cut down either to clear the land for growing crops or for fuel. This was burnt in people's homes, and fuel to make charcoal for industry such as iron, bronze and steel making.

Today the few remaining European forests are preserved for the benefits of the wildlife and people but forests in other parts of the world are being destroyed right now.

In some places such as Nepal, people have no other choice but to cut down their forests for fuel to stay warm, cook their food and boil their water. They depend on the timber to survive.

In tropical forests the cutting and the burning goes on at an even faster rate because wood is used as fuel for industry too.

Wood is not the only fuel, we also burn fossil fuels in our factories, in our cars, to keep our houses warm and our streets lit at night. The amount of energy and the type we use in modern life causes an environmental problem known as global warming.

Air pollution

Industries which burn coal for energy cause air pollution. Forty years ago smoke from these industries mixed with smoke from log fires and fog to form thick smog. This was particularly bad in London where smog was so thick it was hard to see further than a few meters and the air became unpleasant to breathe.

To control this a new law called The Clean Air Act was passed which stopped coal being burned in large towns and cities. As a result the air today is much cleaner and thick smogs are rare.

A London Transport Inspector leads a bus through the smog using a flare. London 1952

Coal smoke releases an element called sulphur into the air. This reacts with rain water to make a very weak acid as strong as vinegar. This is called Acid Rain which damages trees, wildlife and poisons lakes and rivers. Acid rain can be stopped by fitting special gas scrubbers to the chimneys of power stations which clean the smoke and remove the sulphur before it gets into the air.

Global Warming

Fossil fuels such as coal and oil contain carbon. Each time they are burnt some of this carbon combines with oxygen and is released into the air as carbon dioxide gas. Each time anything burns such as coal, or petrol in a car a little more carbon escapes into the atmosphere as carbon dioxide.

The amount of carbon dioxide in the air is quite small. There is less than one in a hundred parts of carbon dioxide in the air compared with larger quantities of other gases like oxygen and nitrogen. But carbon dioxide plays a very important part in the survival of the earth, without it the earth would freeze. It is kept warm by the carbon dioxide which prevents heat escaping from the earth's surface in the same way as the glass in a greenhouse absorbs the heat from the sun. This is sometimes known as the greenhouse effect. The amount of carbon dioxide in the atmosphere has been the same for thousands of years. But extra carbon dioxide has been produced by burning of fossil fuels and the earth has warmed up. This is called global warming. It is our need for energy which is warming up the earth.

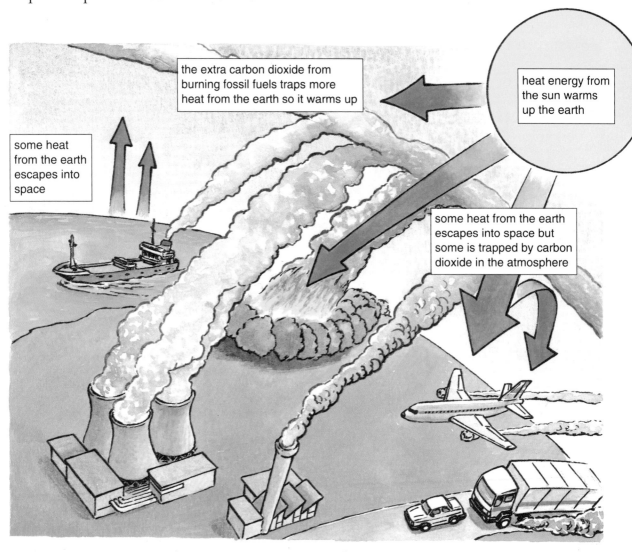

the extra carbon dioxide from burning fossil fuels traps more heat from the earth so it warms up

heat energy from the sun warms up the earth

some heat from the earth escapes into space

some heat from the earth escapes into space but some is trapped by carbon dioxide in the atmosphere

A hotter world may be better for everyone, but nobody can say for certain.

What will the world be like?

A warmer world will heat the oceans. Warm water expands and so the water level will rise. Much of the ice in the polar ice caps will melt adding more water to the oceans. The water level will rise and countries like Holland and Bangladesh which are easily flooded will be flooded completely unless the sea is held back. Cities which are not normally affected by flooding like London and New York will be flooded.

Rainfall will change too so that places which normally have a lot of rain will become drier. Since 1988 each year in Great Britain has been drier than the last although not everyone agrees that this is because of global warming.

These changes will affect the plants and animals which will also change, pests may become more numerous which will cause problems for food growing. Dry countries will become even drier and may suffer longer periods of drought.

Changes happen slowly, but will continue as long as we keep burning fuel to make the energy we need for keeping warm, transport,, growing food and for industry.

We can slow down the process, can you think of ways of doing this; you will get some ideas from PRACTICAL ACTION.

43

What can you do?

You, energy and action

Everything we do uses energy. If we use our own energy it harms the environment less than if we use a fuel. But remember, some of the food we eat may well have been grown using fossil fuels. So even when we use our own energy, the fuel we use (food) may have caused damage to the environment. In our diet we can choose foods which are not produced using fossil fuel energy. For example, less fossil fuel energy is used to produce organic foods. But this means that many organic foods are much more expensive to buy, many people cannot afford them. For others organic food is the only choice because it has not been grown using extra chemicals such as pesticides or artificial fertilisers.

Organic Food

Organic food is grown without having to use extra chemicals such as artificial fertilisers and pesticides. Only natural chemicals from the soil get into organic food. For this reason some people prefer to eat only organic food. Food grown in this way actually saves energy. This is because the energy needed to make artificial chemicals and spray them onto the fields is not used up. Organic foods include not only fruit, vegetables and cereal crops but also livestock such as cattle, chickens and pigs. These are reared in a free-range way, being allowed more space to live and without being fed on extra chemicals that make them grow bigger more quickly.

Processed foods

Processed foods are those which are changed a great deal before we eat them. For example it takes only a small amount of energy to turn potatoes into something we can eat, such as mashed potato, roast potato or chips. Look at the extra energy needed to turn a potato into a packet of crisps.

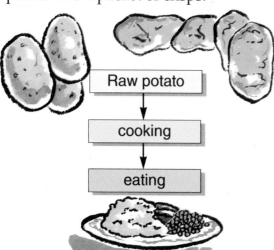

Raw potato

↓

cooking

↓

eating

(Unless you grow your own potatoes extra energy will have been used to get you to the shops in both cases.)

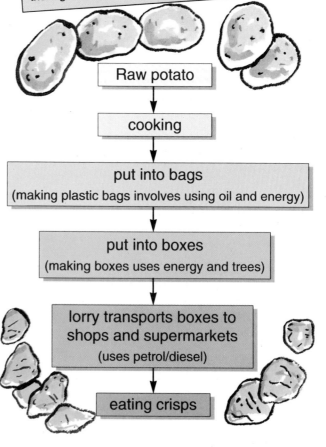

Raw potato

↓

cooking

↓

put into bags
(making plastic bags involves using oil and energy)

↓

put into boxes
(making boxes uses energy and trees)

↓

lorry transports boxes to shops and supermarkets
(uses petrol/diesel)

↓

eating crisps

All those extra stages in processing food use up more energy. Think about all the other foods we buy and the energy which is used to produce them.

What a waste!

Packaging and recycling

Extra energy goes into packaging for food and drink. In today's world we take it for granted, but it is possible to cut down the energy in packaging by recycling. Recycling saves energy because it means that the raw materials used to make packaging do not have to be transported around the World. For example aluminium cans are made from bauxite. This is mined mainly in the tropics. If cans are recycled manufacturers do not need to transport more bauxite from the Tropics, but can simply make new cans by recycling old ones. Glass and paper can also be recycled. The main benefit to the environment is not only reducing the raw materials used, in this case sand and trees, but in the energy saved in manufacturing new glass and paper.

Transport

Instead of going to nearby shops by car encourage your family to walk. Each time the car is used, fossil fuels are burnt adding to global warming. Walking or going by bike is also healthier.

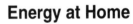

Energy at Home

Keeping lights off when they are not needed, putting on a sweater or warmer clothes, instead of turning up the heat or switching on a fire are just two ways you can help cut down energy use.

Making a compost

Throwing away anything in the dustbin wastes energy. It has to be taken away and burnt or put into a hole in the ground. For some normal household waste such as scraps of leftover food a pet makes a good recycler. If you haven't got a pet you could build a compost heap for food scraps. It's one way of recycling nature's food supply, but also it makes energy available to the 'little rotters' those plants and animals which decompose and rot down nature's waste material so it can be used again by other living things.

Global energy action

Global problems, global solutions

People's needs for energy is greater today than it has ever been before because there are more people in the world than ever before. There is more manufacturing of foods in many different countries of the world. There is more transport of these goods, and the people that buy and sell them and there are more cars

being driven along more roads. All this activity with today's modern machines makes people hungry for energy.

Using energy creates problems for people, places and the whole planet. Nuclear power is thought by some to be a clean, and cheap source of energy. Countries such as France rely on nuclear power stations to supply most of their electricity.

In 1986 in Chernobyl in Russia a nuclear power station exploded, releasing a deadly cloud of radio activity into the air. This poured over much of Europe affecting people, animals and crops. Nuclear power was no longer a safe, clean source of energy. Later, people began to realise that nuclear power was expensive. Closing down a nuclear power station costs a great deal of money to keep it safe. Nuclear power is only a cheap way of producing energy if countries do not have their own coal or oil. In a few years from now there will be a safe, clean and cheap form of nuclear power by a process called nuclear fusion using water as the raw material by a process called nuclear fusion.

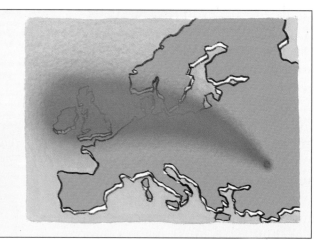

But what is happening now to cut back global problems caused by energy?

Some countries such as Holland are worried by the problem of global warming. Holland is a low, flat country and if global warming continues it will cause a rise in sea level which would flood most of the country. For each new coal-fired power station built in Holland new trees will be planted in tropical forests. How does this help global warming? When coal is burned carbon gets into the atmosphere. The more carbon in the atmosphere the greater the problem of global warming. All plants including trees use carbon to grow, so for every tonne of coal burned to make energy, an equivalent number of trees have to be grown to use up this extra carbon.

Energy saving across the world

Modern ways of building mean that today's buildings conserve energy much better than those built twenty years ago.

In homes

Insulation in the loft and walls

Double glazing

Uses of thermostats on radiators

In the developing world

Bio gas digesters can be used to take plant, animal and human waste and convert it into a gas for fuel. It also produces a solid waste which can be used as fertiliser for soil.

Pump storage power stations

Pumped storage power stations such as at Dinorwig in North Wales operate by pumping water from a low lying lake to a higher lake. The water can be released at any time from the higher lake, rushing down large pipes to spin turbines to produce electricity. Power stations of this type generate electricity at periods of high cost and high demand and pump water back up at periods of low cost and low demand.

Fuelwood plantations

Growing more trees, such as fast-growing Leucaena yields up to 50 tonnes of wood per hectare each year. This is an essential supply for people who use wood as their main source of fuel.

Index

Published by BBC Educational Publishing,
a division of BBC Enterprises Limited,
Woodlands, 80 Wood Lane, London W12 0TT

First published 1993
© Julian Marshall and Steve Pollock/BBC Enterprises Limited 1992

Paperback ISBN: 0 563 35015 6
Hardback ISBN: 0 563 35016 4

Colour reproduction by Daylight Colour, Singapore
Cover origination in England by Dot Gradations
Printed and bound by BPCC Hazell Books, Paulton

Acknowledgments

Photo Credits © A-Z Botanical Collection **page 47 (bottom);** Ancient Art and
Architecture Collection **page 39;** Barnaby's Picture Library **page 11 (top);** British
Museum **page 9;** British Railways Board **page 35 (top);** J Allan Cash Photo Library
page 30; Chubb Racal Security Ltd **page 15 (middle);** Bruce Colman Ltd **pages 7
(top)** A J Deane, **7 (bottom right)** B and C Calhoun, **12, 26 (left)** J Fennell, **28
(middle)** Dr F Sauer, **(bottom)** T O'Keefe; **29** P Wilkinson, **33** H Reinhard, **40** G
Cubitt; Ecoscene **page 21 (bottom)** Wilkinson; Mary Evans Picture Library **pages
10, 21 (top);** Robert Harding Picture Library **pages 24, 26 (right), 37;** Hulton
Deutsch Collection **pages 15 (bottom), 41;** Hutchison Library **page 8,** ICCE **page 45**
C Shayle; Image Bank **page 46** E Hironaka; London Electricity **page 19;** National
Grid Company Plc **page 47 (middle);** National Portrait Gallery, London **page 22;**
Oxford Scientific Films **page 7 (bottom)** D Clyne; Panos Pictures **page 28 (top)** S
Sprague; Science Photo Library **page 27;** Shell Photo Service **page 16;** Still
Pictures **page 47 (top)** M Edwards; Helen Taylor **page 38;** Zefa **pages 2/3** T Ives,
15 (top), 20, 35 (bottom).
Front cover: Science Photo Library; a thermogram of a recently run car.

Illustrations © Mike Atkinson 1992, **pages 4, 5 ,6, 7;** Jon Sayer 1992, **2, 3, 5, 14, 15,
24, 25, 30, 31, 32, 33, 34, 35, 38, 41, 42, 43, 44, 45, 46, 47;** Nick Shewring 1992, **pages
12, 13, 16, 17, 18, 19, 22, 23;** Ross Watton 1992, **pages 8, 11, 20, 21, 26, 27, 29, 36, 37.**